谨以此书献给钢铁工业从业者!

This book is dedicated to people in iron and steel industry

发现钢铁之美

Amazing Steel

本书编委会 编

北 京
冶金工业出版社
2019

图书在版编目（CIP）数据

发现钢铁之美 /《发现钢铁之美》编委会编. —北京：冶金工业出版社，2019.8
ISBN 978-7-5024-8208-4

Ⅰ.①发… Ⅱ.①发… Ⅲ.①钢铁工业—普及读物 Ⅳ.①TF-49

中国版本图书馆 CIP 数据核字（2019）第 176469 号

出 版 人　谭学余

地　　　址　北京市东城区嵩祝院北巷 39 号　邮编　100009　电话　(010)64027926
网　　　址　www.cnmip.com.cn　电子信箱　yjcbs@cnmip.com.cn

责任编辑　曾　媛　美术编辑　彭子赫　版式设计　彭子赫
责任校对　李　娜　责任印制　牛晓波

ISBN 978-7-5024-8208-4

冶金工业出版社出版发行；各地新华书店经销；北京博海升彩色印刷有限公司印刷
2019 年 8 月第 1 版，2019 年 8 月第 1 次印刷
210mm×297mm；10.75 印张；342 千字；163 页
88.00 元

冶金工业出版社　投稿电话　(010) 64027932　投稿信箱　tougao@cnmip.com.cn
冶金工业出版社营销中心　电话　(010) 64044283　传真　(010) 64027893
冶金工业出版社天猫旗舰店　yjgycbs.tmall.com

（本书如有印装质量问题，本社营销中心负责退换）

编委会
EDITORIAL BOARD

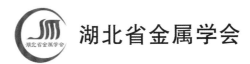 湖北省金属学会

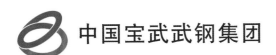 中国宝武武钢集团　　 宝钢股份武钢有限

张寿荣	毛新平	兰　银
蒋扬虎	李艳军	曾　武
李铁林	张　新	李雁宁
林亚萍	杨　帆	顾　钧

序 言
PREFACE

材料是人类文明、社会进步、科学发展的物质基础和技术先导。18世纪钢铁的工业化生产标志着人类由农业社会进入工业社会。相对于其他材料，钢铁更加绿色节能，且100%可循环利用，目前及可预计的将来，由于钢铁材料的经济性，其不可能被大规模替代。

Material is the basis and technological guide of human civilization, social progress and scientific development. The industrialized production of steel in the 18th century marked an agricultural society entering into an industrial society. Compared with other materials, steel is greener and more energy saving, and can be 100% recycled. At present and in the foreseeable future, due its economics, steel can not be replaced on a large scale.

钢铁也号称工业的"粮食"，在社会生产生活的各个领域都有着广泛的应用，我国钢铁工业的发展也是国民经济发展的缩影，从新中国成立之初1949年的18.9万吨到2018年的9.3亿吨，钢铁为我国社会进步做出了重要支撑。

Steel, also known as "food" of other industries, has been widely used in various fields of social production and life. The development of the People's Republic of China's iron and steel industry is also a microcosm of national economic development. From 189,000 tons at the beginning of the founding of the People's Republic of China in 1949 to 930 million tons in 2018, steel has played an important role in China's social progress.

武钢科协（湖北省金属学会）秘书处编辑的《发现钢铁之美》科普图书，将视角聚焦在钢铁构造物上，全面展示钢铁产品对社会多方面的贡献，并介绍了许多钢铁科普知识。

Amazing Steel, a popular science book edited by the Association for Science and Technology of Wuhan Iron&Steel Co., Ltd.（The Hubei Society for Metals）, focuses on steel structures, demonstrating the contributions of steel products to society progress in multiaspect, and introduces a lot of popular science knowledge of steel.

本书采用中英文对照，图文并茂，多视角诠释钢铁与社会、经济的紧密联系，通过精美的照片展示钢铁的美好。作为钢铁行业的科技工作者，我们努力打造更生态的流程，生产更绿色的钢铁，让钢铁与社会共进步。

The book illustrates the beauty of steel as well as the close relationship between steel and society progress from various perspectives through a lot of well-selected photographs with descriptions in Chinese and English. As scientific and technological workers in the steel industry, we strive to create more ecological processes, produce greener steel, and promote the joint progress of steel materials and society.

中国工程院院士
Academician of Chinese Academy of Engineering
宝钢中央研究院副院长
Vice-Director of Baosteel Central Research Institute

毛新平

目 录
CONTENTS

钢铁与城市 ———————————————————————— 1–10
Steel and City

钢铁与工业 ———————————————————————— 11–18
Steel and Industry

钢铁与农业 ———————————————————————— 19–26
Steel and Agriculture

钢铁与工程 ———————————————————————— 27–36
Steel and Engineering

钢铁与建筑 ———————————————————————— 37–46
Steel and Construction

钢铁与交通 ———————————————————————— 47–56
Steel and Transportation

钢铁与桥梁 ———————————————————————— 57–66
Steel and Bridge

钢铁与结构 ———————————————————————— 67–74
Steel and Structure

钢铁与海洋 ———————————————————————— 75–86
Steel and Ocean

钢铁与能源 ———————————————————————— 87–94
Steel and Energy

钢铁与智能 ———————————————————————— 95–104
Steel and Intelligence

钢铁与循环经济 ————————————————————— 105–114
Steel and Recycling Economy

钢铁与生活 ———————————————————————— 115–124
Steel and Life

钢铁与安全 ———————————————————————— 125–136
Steel and Safety

钢铁与健康 ———————————————————————— 137–146
Steel and Health

钢铁与运动 ———————————————————————— 147–154
Steel and Sports

钢铁与艺术 ———————————————————————— 155–162
Steel and Art

后记 ———————————————————————————— 163
Postscript

钢铁与城市
Steel and City

城市是人类文明的产物，钢铁对于现代化城市不可或缺，交通、道路、桥梁、管网、建筑等等，都离不开钢铁，它渗透到我们生活的方方面面。

Cities are the products of human civilization, and steel is the foundation of the modern cities. Steel has penetated into all aspects of the cities where we live from transportation, roads, bridges and pipe networks to buildings, etc.

深圳
Shenzhen

钢铁与城市
Steel and City

钢铁小贴士 Tips on Steel
条材产品 Long Products
线材产品：主要包括帘线钢、电缆钢、钢丝绳用钢等；棒材产品：主要包括易切削钢、建筑钢筋、轴承钢等；型材产品：主要包括重轨、钢板桩等。
Wire products mainly include wire for steel cord, cable steel, wire rope steel and etc.; Bar products mainly include free-cutting steel, reinforcement bar, ball bearing steel and etc.; Section steel products mainly include rail, steel sheet pile and etc.

上海
Shanghai

上海是全球著名的金融中心，世界上规模和面积最大的都会区之一。2010年上海世博会主题：城市让生活更美好。
Shanghai is a world-famous financial center, one of the largest metropolis in population and in area in the world. Theme of the 2010 Shanghai World Expo: Cities make life better.

钢铁与城市
Steel and City

重庆
Chongqing

重庆靠桥梁来跨越山水，成为"中国桥都"，拥有大小桥梁1.3万多座，钢铁连接了天堑。
Chongqing relies on bridges to cross the landscape, becoming a "Chinese Bridge Capital" with more than 13,000 bridges with different sizes.

钢铁与城市
Steel and City

香港是全球第三大金融中心，重要的国际金融、贸易和航运中心，与纽约、伦敦并称为"纽伦港"。全球最高的100栋住宅大楼中，一半位于香港，香港摩天大楼数量居世界首位。

Hong Kong is the third largest financial center in the world, an important international financial, trade and shipping center. It is also known as Newport Harbor together with New York and London. Half of the world's top 100 residential buildings are located in Hong Kong, in which the number of skyscrapers ranks first in the world.

钢铁与城市
Steel and City

香港
Hongkong

钢铁与城市
Steel and City

台北
Taipei

台北是台湾省的经济、文化、教育中心。101大楼以508米的高度成为台北的标志性建筑。
Taipei is the economic, cultural and educational center of Taiwan Province. The 101 building is a landmark in Taipei with a height of 508 meters.

钢铁与城市
Steel and City

东京
Tokyo

东京是全球最大的都市区之一,总人口达3700万。日本主要的钢铁、造船、机械制造、化工、电子和精密仪器等产业都集中在这里,是产业和城市融合的典范。

Tokyo is one of the largest metropolitans in the world with a total population of 37 million. Japan's major industries, such as steel, shipbuilding, machinery manufacturing, chemicals, electronics and precision instruments, are concentrated here, and it is a model for industrial and urban integration.

钢铁与城市
Steel and City

悉尼
Sydney

悉尼位于澳大利亚的东南沿岸，是澳大利亚面积最大、人口最多的城市，被誉为南半球的"纽约"。澳大利亚矿产丰富，我国钢铁行业有40%以上的铁矿石来自这里。

Located on the southeast coast of Australia, Sydney is Australia's largest and most populous city which is known as the "New York" in the Southern Hemisphere. Australia is rich in minerals, and more than 40% of the iron ore in China's steel industry come from here.

钢铁与城市
Steel and City

钢铁小贴士 Tips on Steel

钢的产生可以追溯到3000年前,随着冶炼技术的不断发展进步,今天钢已经成为世界上最具创新性、启发性、多功能性和不可或缺的材料之一。

Discovered more than 3000 years ago, and continuously improved in steel making technology. Today steel has become one of the world's most innovative, inspirational, versatile and essential materials.

云中迪拜
Dubai in the clouds

迪拜,沙漠中崛起的现代化"贸易之都",中东最富有的城市之一,拥有世界上第一家七星级酒店(帆船酒店)、世界最高的摩天大楼(哈利法塔)、全球最大的人工岛(棕榈岛)……这些创造世界纪录的钢铁建筑吸引着全球的目光。

Dubai, the modern "trade capital" rising in the desert, is one of the richest cities in the Middle East, owning the world's first seven-star hotel (sailing hotel), the tallest skyscraper (Halifa Tower) and the largest artificial island (Palm Island) in the world… These world-record steel buildings are attracting global gaze.

钢铁与城市
Steel and City

新加坡
Singapore

新加坡被誉为"亚洲四小龙"之一,是全球第四大国际金融中心,也是亚洲重要的服务和航运中心之一,由于其美丽的市容市貌,被称为"花园城市"。

Known as one of the "Asian Four Little Dragons", Singapore is the fourth largest international financial center in the world and one of the important service and shipping centers in Asia. It is also known as the "Garden City" because of its beautiful city appearance.

钢铁与工业
Steel and Industry

钢铁是工业的粮食。机械、航空航天、轨道交通、船舶、汽车、纺织、家电以及精密仪器等工业门类都离不开钢铁材料作为支撑。

Steel is called "the food" of other industries. Industrial sectors such as machinery, aerospace, rail transit, ships, automobiles, textiles, home appliances, and precision instruments are all inseparable from steel materials.

化工企业
Chemical plant

钢铁与工业
Steel and Industry

汽车制造
Automobile manufacturing

　　钢铁是汽车制造的主要原料，传动系、变速箱、汽车车身都由钢铁打造。目前我国每年汽车行业用钢需求量在5000万吨左右。
　　Steel is the main raw material for automobile manufacturing. The transmission system, gearbox, automobile body and etc. are all made of steel. At present, the demand for steel in the automotive industry is about 50 million tons per year in China.

钢铁小贴士 Tips on Steel

汽车用钢　Automobile steel

　　钢铁是汽车的基本结构材料，其用量占汽车自重的60%~80%，钢板约占50%以上、优质钢（齿轮钢、轴承钢、弹簧钢等特殊钢）占30%、型钢占6%、带钢占6.5%、钢管占3%、金属制品及其他占1%。新制轻型高强度钢（AHSS）有潜力将整个汽车能耗进一步降低50%。

　　Steel is the basic structural material of automobiles, accounting for 60%~80% of the vehicle's own weight, among which, steel plate accounts for more than 50%, high-quality steel (gear steel, bearing steel, spring steel and other special steels) accounts for 30%, section steel accounts for 6%, strip steel accounts for 6.5%, steel pipe accounts for 3%, metal works and others account for 1%. The new light and high strength steel (AHSS) has the potential to further reduce overall vehicle energy consumption by 50%.

钢铁与工业
Steel and Industry

晚霞中的石油开采机
Oil mining machine in sunset

石化工厂
Petrochemical plant

石化工业从勘探、开采、加工、储运，都离不开钢铁。
The petrochemical industry is inseparable from steel for exploration, mining, processing, storage and transportation.

钢铁与工业
Steel and Industry

食品工业
Food industry

不论是食品加工还是食品储存都离不开钢铁。目前很多大型食品加工企业均采用了无人化机械设备生产；由于强度高、耐磨损，全球仅钢制的食品和饮料罐每年就超过2000亿个。

Whether food processing or food storage, it is supported by from steel. At present, many large-scale food processing enterprises have adopted unmanned mechanical equipment production; Due to the its high strength and wear resistance, the food and beverage cans, made by steel only, exceed 200 billion annually in the world.

钢铁与工业
Steel and Industry

航天工业
Aerospace industry

天宫对接
Tiangong docking

我国的航天事业起步于20世纪60年代，截至2018年7月31日，主要的长征系列运载火箭已开发出4代17种型号，飞行282次；在役卫星192颗，居世界第二位；神舟系列飞船和天宫系列空间实验室也达到了世界先进水平。

China's aerospace industry started in the 1960s. By July 31, 2018, the main Long March series carrier rockets have developed 4 generations 17 models and flown 282 times. There are 192 satellites in service, ranking second in the world. The Shenzhou series spacecraft and the Tiangong series space laboratory have also reached the world advanced level.

钢铁与工业
Steel and Industry

盾构机
Shield machine

盾构机是城市地铁、隧道、市政、水电等项目的施工机械，目前我国最大的盾构机直径达15米，长130米，它的出现大大提高了工程进度和效率。

Shield machine is the key construction machinery of urban subway, tunnel, municipal, hydropower and other projects. At present, the largest shield machine in China has a diameter of 15 meters and a length of 130 meters which greatly improved the progress and efficiency of the construction projects.

钢铁与工业
Steel and Industry

钢铁小贴士 Tips on Steel

特种钢 Special steel

特种钢也叫合金钢，是一种钢材。特种钢是在碳素钢里适量地加入一种或几种合金元素，使钢的组织结构发生变化，从而使钢具有各种不同的特殊性能。第二次世界大战后，世界各军事强国为了满足舰船装备的发展需求，研制开发了系列高强度舰船用钢，这就是特种钢的早期雏形。

Special steel, also known as alloy steel, is a kind of steel with one or several alloying elements added into carbon steel appropriately which change the structure of steel to achieve a variety of special properties. After World War II, in order to meet the needs of naval equipment development, the military powers in the world began to develop high strength naval steels, which are the early forms of special steel.

精密加工
Precision machining

随着现代机械加工的快速发展，涌现出了许多先进的机械加工技术方法，例如微型机械加工技术、快速成形技术、超精密加工技术等。当然精密加工的钢材都是特种钢材。

With the rapid development of modern machining, many advanced machining techniques, such as micromachining technology, rapid prototyping technology, ultra-precision machining technology, etc., have emerged. Of course, the object of precision processing is special steel.

钢铁与工业
Steel and Industry

雪龙2号
XUE LONG 2

2018年9月下水的"雪龙2号"极地考察船,是我国自己建造的首艘极地破冰船,且为全球第一艘采用船艏、船艉双向破冰技术的极地科考破冰船。该船可突破极区20米冰脊,续航里程2万公里,满足全球无限航区航行需求。

"XUE LONG 2", launched in September 2018, is the first polar icebreaker made by China. It is the world's first polar scientific icebreaker with two-way icebreaking technology. The ship can break through the 20-meter ice ridge in the polar region and has a cruising range of 20,000 kilometers, meeting the navigation needs of the global unlimited navigation area.

钢铁与农业
Steel and Agriculture

农业是现代文明的基础。它给我们提供衣食，并在越来越大的程度上提供能源。农业为全球提供35%的从业机会。从基本的锄头、铁锹和耙子，到现代化的耕具、灌溉系统，每个步骤都有钢铁的影子，它让农业变得更容易、更有效率。

Agriculture is the foundation of modern civilization. It provides us with food, clothing and also the energy to larger extent. Agriculture provides 35% of the world's employment opportunities. From hoes, shovels and rakes to modern farming tools and irrigation systems, steel is used in every step, which makes agriculture easier and more efficient.

水肥一体滴灌
Drip irrigation with water and fertilizer

钢铁与农业
Steel and Agriculture

收割机
Harvester

大型收割机降低了农业劳动强度,实现高速批量的生产。
Large harvesters reduce the intensity of agricultural labor and achieve high-speed batch production.

钢铁与农业
Steel and Agriculture

奶牛养殖场
Dairy farm

钢也是自动化畜牧养殖系统的理想材料,包括牲畜通过用钢铁制造的卡车、火车、飞机运送到全球各地市场。
Steel is also an ideal material for automated animal aquaculture systems. Livestock are transported to markets around the world by trucks, trains and airplanes, which are all made of steel.

钢铁与农业
Steel and Agriculture

农用拖拉机
Agricultural tractors

从基础的农业工具，例如锹、镐、犁，到技术先进的拖拉机、犁地机、收割机，钢铁无处不在。
From basic agricultural tools such as shovels, picks and ploughs to advanced tractors, ploughs and harvesters, steel is everywhere.

钢铁与农业
Steel and Agriculture

农业大棚
Agricultural greenhouses

大型批量化生产的农业作棚，均使用易拼接、轻量防腐蚀的钢结构搭建而成，它保护植物在正常的温度下生长，提高产量。
Large-scale and mass-produced agricultural greenhouses are constructed with light weight corrosion-resistant steel structures, which can ensure plants growing at normal temperatures and increase production.

钢铁与农业
Steel and Agriculture

粮食仓库
Granary

钢铁建成的粮仓，可以保护粮食不受虫子、老鼠等的侵食。
A granary built of steel can protect food from insects, rats, and etc. effectively.

钢铁与农业
Steel and Agriculture

钢铁小贴士 Tips on Steel

铁是地球排名第4位的常见元素，位列氧（46%）、硅（28%）和铝（8%）之后。当铁水转换成钢时，温度高达1700℃，远高于火山熔岩的温度。

Steel is the 4th most common element in the Earth's crust after oxygen (46%), silicon (28%), and aluminum (8%). When liquid iron is converted into steel it reaches temperatures of up to 1700℃, significantly hotter than volcanic lava.

当铁与碳、废钢以及少量其他元素结合时，就会转换成一种强度更高的材料——钢，成为人民生活中最广泛使用的材料。钢是铁与碳的合金，含有低于2%的碳和1%的锰以及少量硅、磷、硫和氧。

When iron is combined with carbon, recycled steel and small amounts of other elements, it is transformed into a much stronger material called steel, used in a huge range of human-made applications. Steel is an alloy of iron and carbon, containing less than 2% carbon, 1% manganese and small amounts of silicon, phosphorus, sulphur and oxygen.

铁犁
Iron plough

公元前6世纪，我国就发明了铁犁，有的地方至今还在使用。在没有机械化耕种之前，它的出现大大降低了农民的劳动强度，提高了农业产量。

In the 6th century BC, iron plough was invented in China, and it is still in use in some places in the earth. Before mechanized tillage, it greatly reduced the labor intensity of farmers and increased agricultural output.

钢铁与农业
Steel and Agriculture

五彩的农田
Colorful farmland

钢渣作为碱性渣,其中的氧化钙、氧化锰可以改善酸性土壤的土质。含高磷的钢渣可用于缺磷的碱性土壤中并增强农作物抗病虫害的能力。硅是水稻生长需求最大的元素,二氧化硅含量高于15%的钢渣可以作硅肥。

As alkaline slag, calcium oxide and manganese oxide in steel slag can improve the soil quality of acidic soil. Steel slag with high phosphorus content can be used in alkaline soil lacking phosphorus and enhance crop's resistance to diseases and pests. Silicon is the most important element for rice growth. Steel slag with more than 15% silicon dioxide content can be used as silicon fertilizer.

钢铁与工程
Steel and Engineering

钢铁由于其高强度、耐久性、可循环、多功能及经济性等于一体的独特特征，是各种工程不可或缺的基础材料。

Steel is an indispensable foundation material for various projects due to its unique characteristics of high strength, durability, recyclability, versatility and economy.

工程挖掘机
Engineering excavator

钢铁与工程
Steel and Engineering

京沪高速铁路
Beijing-Shanghai high-speed railway

全球第一次建成里程最高的高速铁路，全长1318公里，时速350公里，连接北京、上海两个超级大都会和天津、济南、南京、无锡、苏州等东部地区重要城市，使京沪两市的陆路交通时间自此缩短到5小时以内。

The world's first longest high-speed railway, with a total length of 1,318 kilometers and a speed of 350 kilometers per hour, connects two super-metropolitan cities, Beijing and Shanghai, and other important cities in the eastern regions such as Tianjin, Jinan, Nanjing, Wuxi and Suzhou, shortening the land transportation time between Beijing and Shanghai to less than 5 hours.

钢铁与工程
Steel and Engineering

杭州湾跨海大桥
Hangzhou bay sea cross bridge

　　杭州湾的跨海大桥2008年竣工时是当时中国最长的跨海大桥。大桥南起宁波慈溪，北到嘉兴海盐，使上海到宁波之间无需再绕行杭州，缩短了沪甬之间约120公里的行车距离。该大桥全长36公里，所用钢量达80万吨，相当于7个鸟巢体育馆的用钢量。
　　The Hangzhou Bay Sea Cross Bridge was the longest sea-crossing bridge in China at the time of its completion in 2008. The bridge, starting from Cixi, Ningbo in the south, and to Jiaxing, Haiyan in the north, makes it unnecessary to bypass Hangzhou from Shanghai to Ningbo, shortening the distance between Shanghai and Ningbo by about 120 kilometers. The bridge is 36 kilometers long and 800,000 tons of steel was used, which is equivalent to the amount of steel of seven Bird's Nest stadiums in Beijing.

钢铁与工程
Steel and Engineering

南水北调工程
South-to-North water diversion project

"南水北调"是全球最大的调水工程。其中,中线工程将汉水上游丹江口水库的水调往京津冀豫四省,为沿线十几座大中城市提供生产生活和工农业用水。供水范围内总面积15.5万平方千米,输水干渠总长1277公里。

"South-to-North Water Diversion" is the world's largest water transfer project. Among them, the mid-line project transfers the water from the Danjiangkou Reservoir of upper Hanshui to the four provinces of Beijing, Tianjin, Hebei and Henan, providing production, living, industrial and agricultural water for more than a dozen large and medium-sized cities along the line. The total area of the water supply is 155,000 square kilometers, and the total length of the water mains is 1277 kilometers.

钢铁与工程
Steel and Engineering

西气东输工程
West-to-East gas transmission

"西气东输"是我国距离最长、口径最大的输气管道,西起塔里木盆地的轮南,东至上海。自新疆轮台县塔里木轮南油气田,东西横贯新疆、甘肃、宁夏、陕西、山西、河南、安徽、江苏、上海等9个省区,全长4200千米,用钢200余万吨。

"West-to-East Gas Transmission" is the longest in length and largest in diameter gas pipeline in China, starting from Lunnan in the Tarim Basin in the west to Shanghai in the east. From the Tarim South Oil and Gas Field in Luntai County, Xinjiang, it passes through nine provinces and autonomous regions from the east to the west, Xinjiang, Gansu, Ningxia, Shanxi, Shanxi, Henan, Anhui, Jiangsu and Shanghai, with a total length of 4,200 kilometers and more than 2 million tons of steel.

钢铁小贴士 Tips on Steel

在石油和天然气行业,从钻探和抽取到加工和分配,钢是最核心的材料。运输管道必须在极为苛刻的环境条件下满足特定需要。

Steel is essential in the oil and gas industry, from drilling and extraction to processing and distribution. Pipes must meet specific needs in highly challenging environments.

钢铁与工程
Steel and Engineering

天眼——贵州500米口径球面射电望远镜（简称FAST）
Five-hundred-meter aperture spherical radio telescope

　　FAST的口径有500米，近30个足球场大的接收面积，主反射面的面积达25万平方米，是具有我国自主知识产权、世界最大单口径、最灵敏的射电望远镜，其设计综合体现了我国高技术创新能力。
　　FAST, 500 meters of caliber, with a large receiving area of nearly 30 football fields, and the main reflecting surface covers an area of 250,000 square meters, is the most sensitive and the world's largest single-caliber radio telescope with Chinese independent intellectual property rights. It reflects China's high-tech innovation capability.

钢铁与工程
Steel and Engineering

三峡工程
Three gorges project

  三峡大坝用了192万吨钢材，1,082万吨水泥，建成高185米的大坝，蓄水高度175米，安装32台单机容量为70万千瓦的水电机组，2018年发电量突破千亿千瓦时，是全球最大的水利枢纽工程。宝武集团高磁感取向硅钢应用于三峡地下电站变压器上，打破了取向硅钢依靠进口的局面。

  1.92 million tons of steel and 10.82 million tons of cement were used in Three Gorges Project to build a 185-meter-high dam with a water storage height of 175 meters. It installed 32 hydropower units with a capacity of 700,000 kilowatts per unit. In 2018, the power generation exceeded 100 billion kilowatts hours, making it the largest water conservancy project in the world at that time. Baowu Group's high magnetic induction oriented silicon steel was applied to the transformer of the Three Gorges underground power station, breaking the situation that the oriented silicon steel relied on imports.

> **钢铁小贴士 Tips on Steel**
>
> **电工钢 Electrical steel**
>
>   电工钢分无取向和取向电工钢。无取向电工钢，顾名思义，其磁性具有各向同性（也称各向同性电工钢），广泛应用于制造各类电机、电子变压器铁芯等。在我国，无取向电工钢的使用量约占整个电工钢使用量的90%。取向硅钢是晶粒的易磁化方向平行于轧向、含硅量3%左右的电工钢，多用于电能传输领域（如变压器等），是电力、电子和军事工业不可缺少的重要金属功能材料。
>
>   Electrical steels are divided into non-oriented and oriented electrical steels. Non-oriented electrical steel, as its name implies, has isotropic magnetism (also known as isotropic electrical steel), which is widely used in manufacturing various types of motor and electronic transformer cores. In China, the usage of non-oriented electrical steel accounts for about 90% of the total usage of electrical steel. Oriented silicon steel is an electrical steel whose grain magnetization direction is parallel to rolling direction and silicon content is about 3%. It is mostly used in the field of power transmission (such as transformer) and is an indispensable important metal functional material for power, electronics and military industries.

钢铁与工程
Steel and Engineering

亚欧铁路东起中国的连云港、日照等沿海港口城市,西行出域穿越哈萨克斯坦等中亚地区,经俄罗斯、白俄罗斯、乌克兰、波兰、德国、法国、西班牙、荷兰等欧洲口岸,全程超过13000公里左右,是不折不扣的"国际钢铁走廊"。

The Asia-Europe Railway, more than 13,000 kilometers, starts from China's coastal port cities such as Lianyungang and Rizhao, and westward to Kazakhstan and other Central Asian regions, via many European ports, such as Russia, Belarus, Ukraine, Poland, Germany, France, Spain, the Netherlands and etc. It is an uncompromising "International Steel Corridor".

亚欧铁路
Asia-Europe railway

钢铁与工程
Steel and Engineering

C919客机
C919 aircraft

2017年5月5日首飞的C919大飞机打破了欧美的垄断。起落架是支撑飞机的唯一部件，只有特种钢才能满足其对材料强度、韧性等的要求，C919的起落架已经实现国产化。

The C919 big plane, flew for the first time on May 5, 2017, broke the monopoly of Europe and the United States. The landing gear is the only component that supports the aircraft. Only special steel can meet its requirements for the material strength and toughness. The landing gear of the C919 has been localized in China.

钢铁与工程
Steel and Engineering

上海世博会中国馆
China pavilion at shanghai World Expo

中国馆用钢结构撑起了层叠出挑，凝聚中国元素、象征中国精神的宝鼎结构，使其具有极大的震撼力和视觉冲击力。
The structure of layer upon layer is propped up with steel structure in China Pavilion , which embodies the Chinese elements and symbolizes the Chinese spirit, making it extremely powerful and visually impactful.

钢铁与建筑
Steel and Construction

目前，世界年产钢的一半用在了基础设施和建筑上，钢铁以其经济实惠、容易获得，且集强度、多功能性、耐用性和可循环性等众多优点为设计师及建筑施工提供解决方案。

At present, half of the world's annual steel is used in infrastructure and construction. Because of its many advantages, such as economical and easy to obtain, and its strength, versatility, durability and recyclability, steel provides a solution for constructions.

珠海大剧院
Zhuhai grand theatre

钢铁与建筑
Steel and Construction

阿联酋哈利法塔
Burj Khalifa Tower, United Arab Emirates

哈利法塔是目前世界第一高楼，高828米，楼层总数162层，造价15亿美元。哈利法塔总共使用6.2万吨强化钢筋用以支撑33万立方米混凝土和14.2万平方米玻璃。大厦内设有56部电梯，速度最高达17.4米/秒。

Burj Khalifa Tower is currently the world's tallest building, 828 meters high, with 162 floors and a cost of $1.5 billion. The Burj Khalifa used a total of 62,000 tons of reinforced steel to support 330,000 cubic meters of concrete and 142,000 square meters of glass. There are 56 elevators in the building with speeds up to 17.4 m/s.

钢铁与建筑
Steel and Construction

北京首都国际机场3号航站楼用钢50多万吨，像一个展翅的大鹏。年运输旅客可达8600万人，是世界上最繁忙的机场之一。

Beijing Capital International Airport Terminal 3, with an annual transportation capacity of 86 million passengers, is one of the busiest airports in the world. It used more than 500,000 tons of steel, and looks like a winged big bird.

首都国际机场3号航站楼
Terminal 3 of capital international airport

钢铁小贴士 Tips on Steel

建筑钢 Construction steel

建筑用钢具有碳当量低、焊接性能好、强度高、屈强比低的特点。钢铁用于建筑的潜力无限，在所有民用建筑中，钢材的强重比是最高的，钢制框架、结构梁和地基能很好地支撑房子，钢制板和屋顶能抵御极端的气候变化、遮风挡雨。

Steel for construction has the characteristics of low carbon equivalent, good welding performance, high strength and low yield ratio. Its potential advantage for construction is unlimited. In all civil buildings, steel has the highest strength-to-weight ratio. Steel frames, structural beams and foundations support the house well. Steel panels and roofs resist extreme climate change, keep out the wind and rain.

钢铁与建筑
Steel and Construction

中央电视台
China central television(CCTV)

  中央电视台总部大楼设计对传统高楼的建筑提出了挑战,也打破了北京的建筑代码。它由两个斜塔组成,每一个都倾斜90度,形成一个连环。大楼高234米,用钢量14万吨。
  The design of the CCTV building headquarters challenged the construction of traditional high-rise buildings and broke the building code of Beijing. It consists of two leaning towers, each inclined at 90 degrees to form a chain. The building is 234 meters high and has a steel consumption of 140,000 tons.

钢铁与建筑
Steel and Construction

重建的纽约世贸中心
Reconstructed New York world trade center

世界贸易中心一号楼（自由塔），钢结构，号称纽约最环保的建筑之一。坐落于"911"袭击事件中倒塌的原世界贸易中心的旧址。高度541.3米，1776英尺，1776年为独立宣言发布年份。

World Trade Center building 1 (Freedom Tower), with steel structure, is one of the most environmentally friendly buildings in New York. It located on the site of the former World Trade Center, which collapsed during the September 11 attacks in 2001. It is 541.3 meters high, or 1776 feet, and 1776 was the year of the establishment Declaration of Indepence.

钢铁与建筑
Steel and Construction

> **钢铁小贴士** Tips on Steel
>
> 在众多建筑材料中，钢的强度/重量比最高，可降低建筑物的重量，因此地基基础埋置深度不会太深，建造成本也更低。
> Steel offers the most economic and the highest strength to weight ratio of any building material, resulting in lighter buildings requiring less extensive and costly foundations.
>
> 因为有钢，摩天大楼才成为可能。2017年，全球高度在200米以上的建筑物总数为1319座，而在2000年只有263座，增长了402%。
> Skyscrapers are made possible by steel. In 2017, the total number of buildings in the world over 200 meters high was 1319, a 402% increase from the year 2000, when the number was just 263.

吉隆坡石油双塔
Kuala lumpur petronas twin towers

吉隆坡石油双塔于1997年建成，高452米，地上88层，曾经是世界最高的摩天大楼，目前仍是世界最高的双塔楼。大楼表面大量使用了不锈钢材质，现代感十足，是吉隆坡的知名地标及象征。

Built in 1997, 452 meters high and 88 stories above ground, the Petronas Twin Towers is a well-known landmark and symbol of Kuala Lumpur. It was once the tallest skyscraper in the world and is still the tallest in the world as a twin tower now. Using mass of stainless steel at the surface of the building made it modern.

钢铁与建筑
Steel and Construction

集装箱房屋是一种又一次撞击时尚潮流的建筑体系，可随时随地移动各地，为人们带来更方便更舒适的生活。
Container house is a kind of building series, which has hit the fashion trend again and again. It can be moved anytime and anywhere to bring people more convenient and comfortable life.

集装箱房屋
Container house

钢铁与建筑
Steel and Construction

黄鹤楼
Yellow crane tower

黄鹤楼始建于三国时代吴黄武二年（公元223年），有"天下江山第一楼"之称。由于之前是木质结构，几经烧毁。现在这座黄鹤楼是1981重建的钢混结构，楼高5层。

The Yellow Crane Tower was firstly built in the second year of Wu Huangwu (AD 223) in the Three Kingdoms Period and is known as the "first floor of the world". Due to the previous wooden structure, it was burned several times. Now the Yellow Crane Tower is a steel–concrete structure rebuilt in 1981 with five floors.

钢铁与交通
Steel and Transportation

飞机和高铁
Aircraft and high-speed railway

　　钢铁是汽车、飞机、轮船、高铁、桥梁等的最主要的工程材料。钢铁为人类提供更快速、高效、经济的交通出行方案。全球大约16%的钢铁被用于满足社会的交通运输需求。在其他相关基础设施中，钢铁也必不可少，例如，公路、港口、车站、机场、加油站等。

　　Steel is the most important engineering material for automobiles, airplanes, ships, high-speed rails, bridges and etc. Steel provides a more rapid, efficient and economical transportation plan for human beings. About 16% of the world's steel is used to meet the transportation needs of society. In other related infrastructure, steel is also indispensable, such as roads, ports, stations, airports, gas stations and so on.

钢铁与交通
Steel and Transportation

威尼斯的刚朵拉
Gondola in Venice

水城威尼斯的交通工具就是刚朵拉。
Gondola, Venice's water taxi, is the main transportation in this watercity.

钢铁与交通
Steel and Transportation

钢铁小贴士 Tips on Steel
钢轨产品 Rail products

钢轨是轨道结构的重要部件,起到承担载荷、引导车轮的作用,在电气化铁路上,钢轨还兼做轨道电路,传递信号。高速和重载铁路的快速发展,对钢轨重量提出了更高的要求,钢轨要为车轮提供连续、平顺和阻力最小的滚动平面。我国铁路建设持续快速发展,目前钢轨年需求量约为300万吨,主要以50kg/m、60kg/m和75kg/m为主,材质主要包括U75V和U71Mn等珠光体钢轨,产品广泛应用于国家铁路工程和城市轨道交通工程。

Rail is an important component of track structure, which plays the role of bearing load and guiding wheel. On the electrified railway, the rail also serves as a track circuit to transmit signals. Rapid development of high-speed and heavy-duty railways puts higher demands on rail weight. The rails provide a continuous, smooth, and least resistant rolling plane for the wheels. China's railway construction continues to develop rapidly. At present, the annual demand for rails is about 3 million tons, mainly 50kg/m, 60kg/m and 75kg/m. The materials mainly include pearlite rails such as U75V and U71Mn. The products are widely used in national railway engineering and urban rail transit engineering.

地铁
Subway

我国到2020年城市轨道交通将达7500公里,城市轨道建设年用钢需求量在4000万吨左右。
By 2020, China's urban rail transit will reach 7,500 kilometers in length, and the annual demand for steel for urban rail construction will be about 40 million tons.

钢铁与交通
Steel and Transportation

夜幕下的公交车站
Bus stops at night

钢铁小贴士 Tips on Steel

新开发的轻型高强度钢使所应用的产品可实现25%~40%减重及强度增加。汽车制造商采用先进高强度钢,打造更加安全的汽车。有些等级的先进高强度钢经过工程设计,能够吸收前部撞击等产生的撞击能量,还有些等级的先进高强度钢可以偏转侧面撞击等产生的撞击能量。

New lightweight high-strength steels can make their applications 25%~40% lighter & stronger. Car manufacturers use Advanced High-Strength Steel (AHSS) for safer vehicles. Some AHSS grades are engineered to absorb crash energy, such as from front crash, and some are engineered to deflect crash energy such as from side crash.

当我们采用全生命周期评价方法,比较功能对等的汽车组件时,先进高强度钢的性能一直领先于低密度竞争材料,并且在生产阶段排出的二氧化碳量,比铝或碳钎维低5倍,比镁低7倍。

When taking a life cycle approach to compare functionally equivalent automotive components, AHSS consistently outperform lower density competing materials, emitting in the production phase 5 times less CO_2 than aluminium or carbon fibre, and 7 times less CO_2 than magnesium.

钢铁与交通
Steel and Transportation

兰卡威群岛天空之桥
Sky bridge over the lankawi islands

钢铁小贴士 Tips on Steel

交通网络：桥梁、隧道、铁轨以及加油站、火车站等基础设施都需要钢铁。
Transportation network: Steel is needed for infrastructure such as bridges, tunnels, rails, gas stations and railway stations.

钢铁与交通
Steel and Transportation

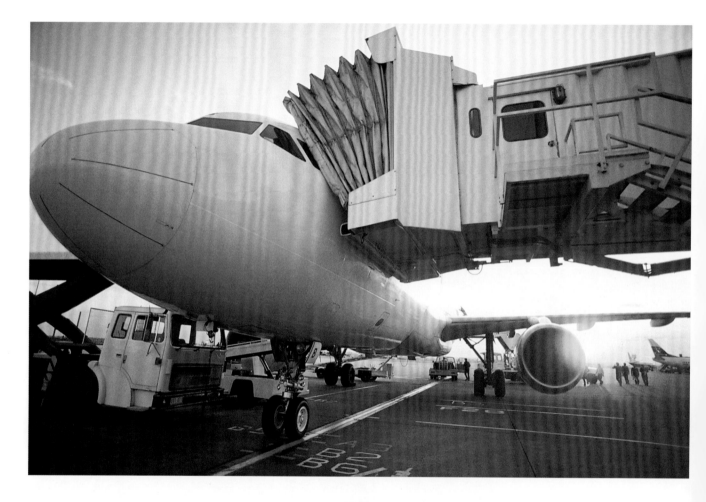

飞机
Aircraft

超高强度的特钢使用在飞机起落架上。没有钢铁,飞机将无法完成安全的起降。
Super-strength special steel is used in aircraft landing gear. Without steel, airplanes would not be able to take off and land safely.

钢铁与交通
Steel and Transportation

高速公路
Expressway

我国高速公路，总里程全球第一。为保证高速公路的耐用性和安全性，每公里高速公路使用钢材约为450吨。
The total mileage of expressway in China ranks first in the world. In order to ensure the durability and safety of expressways, about 450 tons of steel are used per kilometer of expressways.

钢铁与交通
Steel and Transportation

　　铁路运输中的火车、铁轨和基础设施都需要钢铁。在中短途运输上，与几乎所有其他运输方式相比，铁路减少了乘客单位公里的在途时间和二氧化碳排放量。

　　Steel is required for trains, rails and infrastructure in rail transport. Compared with almost all other types of transport, railways reduce the travel time per kilometre of passengers and carbon dioxide emissions.

和谐号高铁
Harmony high-speed rail

钢铁与交通
Steel and Transportation

钢铁小贴士 Tips on Steel

在高速火车中，钢铁占其质量的20%~25%。火车的主要钢质组件是转向架（位于火车下方的结构，包括车轮、车轴、轴承和马达）。货车车厢几乎完全由钢铁制成。

In high-speed trains, steel accounts for 20%~25% of its weight. The main steel component of the train is the bogie (structure under the train, including wheels, axles, bearings and motors). The truck compartment is almost entirely made of steel.

今天，全球铁路网的长度是1,051,767千米。这相当于绕地球26圈。

Today, the length of the worldwide rail network spans 1,051,767 km. This is equivalent to going round the earth 26 times.

钢铁与交通
Steel and Transportation

香港有轨电车
Trams in Hong Kong

钢铁与桥梁
Steel and Bridge

钢铁是现代桥梁的核心组成部分，钢铁强度高，可以弯曲又不易折断，即使在严酷的条件下，也拥有较长的使用寿命。

Steel is the core component of modern bridges. Besides its high-strength, steel also can be bent and is not easily broken. It has a long service life even under severe conditions as well.

武汉天兴洲铁路公路两用桥
TianXingZhou railway and road dual-purpose bridge in Wuhan

钢铁与桥梁
Steel and Bridge

世界第一座铁桥
The world's first iron bridge

　　1779年在塞文河上建起的世界第一座铁桥位于什鲁斯伯里（Shrewsbury），其作为工业革命的发源地而闻名于世。这座铁桥跨度100英尺，高52英尺，宽18英尺，全部用铁浇铸，重384吨。
　　The world's first iron bridge, built on the Saiwen River in 1779, located in Shrewsbury, is famous as the birthplace of the industrial revolution. The bridge, 52 feet tall and 18 feet wide, was all cast in iron, which spans one hundred feet and weighs 384 tons.

钢铁与桥梁
Steel and Bridge

钢铁小贴士 Tips on Steel
桥梁钢 Bridge steel

桥梁钢板是专用于架造铁路或公路桥梁的钢板。有较高的强度、韧性以及承受机车车辆的载荷和冲击，且要有良好的抗疲劳性、一定的低温韧性和耐大气腐蚀性。与传统钢材相比，高强度的桥梁钢可用于建造成本更低、强度更大、耐候性更好的桥梁，且比传统材料设计桥梁轻28%。

Bridge steel plate, with high strength and toughness, is used to build railway or highway bridges. It has good fatigue resistance, low temperature toughness and atmospheric corrosion resistance. Compared with traditional steel, high strength bridge steel can be used to build bridges with lower cost, higher strength and better weatherability, which is 28% lighter than traditional material design bridges.

胶州湾大桥
Jiaozhou bay bridge

胶州湾大桥全长36.48公里，连接着青岛和黄岛。双向六车道由5200根钢柱支撑，总共用钢45万吨，桥梁能够承受高达8.0级的地震。

The Jiaozhou Bay Bridge, with a total length of 36.48 kilometers, connects Qingdao and Huangdao. The two-way six-lane bridge is supported by 5200 steel columns, with a total of 450,000 tons of steel. The bridge can withstand earthquakes of magnitude 8.0.

钢铁与桥梁
Steel and Bridge

伦敦塔桥
Tower bridge

　　伦敦塔桥两端由4座石塔连接，两座主塔高43.455米。河中的两座桥基高7.6米，相距76米。桥塔和桥身用了1.1万吨钢铁，骨架外铺设花岗岩和波特兰石来保护骨架和增加美观。
　　The Tower Bridge in London is connected by four stone towers, the two main towers of which are 43.455 meters high. The other two bridge foundations in the river are 7.6 meters high and 76 meters apart. The pylon and body of the bridge are made of 11,000 tons of steel. Granite and Portland stone are laid outside the skeleton to protect the skeleton and enhance its beauty.

钢铁与桥梁
Steel and Bridge

北盘江大桥全长1341.4米,最大跨径720米,桥面距江面垂直高度达565.4米,相当于200层楼高,2018年被世界吉尼斯纪录认证为"世界最高桥"称号。

Beipanjiang Bridge is 1341.4 meters long, with a maximum span of 720 meters. The vertical height is 565.4 meters, which is equivalent to the height of 200 stories. It was certified as "the highest bridge in the world" by the Guinness World Records in 2018.

北盘江大桥
Beipanjiang bridge

钢铁与桥梁
Steel and Bridge

重庆朝天门长江大桥
Chaotianmen Yangtze river bridge in Chongqing

朝天门大桥有两座主墩，主跨达552米，比世界著名拱桥——澳大利亚悉尼大桥的主跨还要长，成为"世界第一拱桥"。
The Chaotianmen Bridge has two main piers with a main span of 552 meters, which is longer than the main span of the world famous arch bridge, the Sydney Bridge in Australia, and has become the "world's first arch bridge".

钢铁与桥梁
Steel and Bridge

上海外白渡桥
Wai Bai Du bridge in Shanghai

外白渡桥是中国的第一座全钢结构铆接桥梁，于1908年1月20日落成通车。由于其悠久的历史和独特的设计，外白渡桥成为上海的标志之一，同时也是上海的现代化和工业化的象征。

Wai Bai Du bridge, the first full steel riveting bridge in China, was opened in January 20, 1908. Due to its long history and unique design, Wai Bai Du bridge has become one of the symbols of Shanghai, and is also a symbol of Shanghai's modernization and industrialization.

钢铁与桥梁
Steel and Bridge

金门大桥
Golden gate bridge

钢铁与桥梁
Steel and Bridge

金门大桥峙于美国加利福尼亚州旧金山金门海峡之上，是世界著名的桥梁之一，桥身全长1900多米。历时4年，利用10万多吨钢材，耗资达3550万美元建成。

Standing on the Golden Gate Strait in San Francisco, California, USA, the Golden Gate Bridge, with 1900 meters in length, is one of the world's famous bridges. It was built for four years with more than 100,000 tons of steel and costed 35.5 million US dollars.

钢铁与桥梁
Steel and Bridge

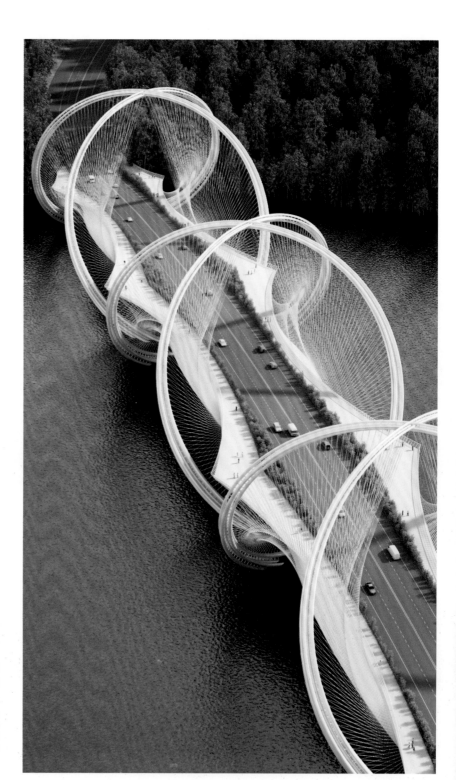

三山大桥
Sanshan bridge

三山大桥的设计灵感源自奥运五环。全长452米的大桥横跨妫河两岸，由三组交叉项链的拱形钢结构组成，最大跨度达95米。

The design of Sanshan Bridge was inspired by the Olympic rings. The 452-metre-long bridge spans both sides of the Gui River and consists of three groups of arch steel structures with crossed necklaces, with a maximum span of 95 metres.

钢铁与结构
Steel and Structure

钢铁小贴士 Tips on Steel
彩涂板 Color coated sheet

彩涂板,又称彩钢板,彩板。彩色涂层钢板是以冷轧钢板和镀锌钢板为基板,经过表面预处理以连续的方法涂上涂料,经过烘烤和冷却而制成的产品。涂层钢板具有轻质、美观和良好的防腐蚀性能,又可直接加工。它给建筑业、造船业、车辆制造业、家电行业、电气行业等提供了一种新型原材料,代表了未来绿色环保新材料发展的方向。

Color coated sheet, also known as color steel sheet or color sheet, is a product obtained by subjecting a cold-rolled steel sheet and a galvanized steel sheet to a substrate, a surface pretreatment, a coating method in a continuous way, and baking and cooling. With light weight, beautiful appearance and good anti-corrosion performance, and direct processing, it provides a new type of raw material for the construction industry, shipbuilding industry, vehicle manufacturing industry, home appliance industry, electrical industry, and etc., representing the future of green environmental protection.

武汉光谷科技城
Guanggu science and technology city in Wuhan

钢铁与结构
Steel and Structure

北京奥林匹克塔
Beijing Olympic tower

北京奥林匹克塔由五座186米至246.8米高的独立塔组合而成，其以"生命之树"为设计理念，寓意为生命之树从地壳破土而出，五个高低不同的钢结构塔身在空中似合似分，造型独特，蕴含着奥运五环蓬勃向上的精神风貌。

The Beijing Olympic Tower is composed of five independent towers, from 186 meters to 246.8 meters high. The design concept, "Tree of Life", implies the tree of life emerging from the earth's crust. The five steel structure towers of different heights form unique shapes in the air, containing the spiritual features of the five Olympic rings.

钢铁与结构
Steel and Structure

武汉青山江滩钢结构造型

Steel structural modeling of Qingshan river beach in Wuhan

钢铁与结构
Steel and Structure

广州电视塔

Guangzhou TV tower

广州电视塔塔身主体高454米，天线桅杆高146米，总高度600米。其造型、空间和结构由两个向上旋转的椭圆形钢外壳变化生成，呈现"纤纤细腰"，所以有个别号叫"小蛮腰"。

The total height of the Guangzhou TV Tower is 600 meters, with the tower body height of 454 meters, and the antenna mast height of 146 meters. Two upward rotating oval steel shell make its apperance, showing a "slender waist" of the tower, so it is also called "Xiao Man waist".

钢铁与结构
Steel and Structure

温哥华街景
Vancouver street view

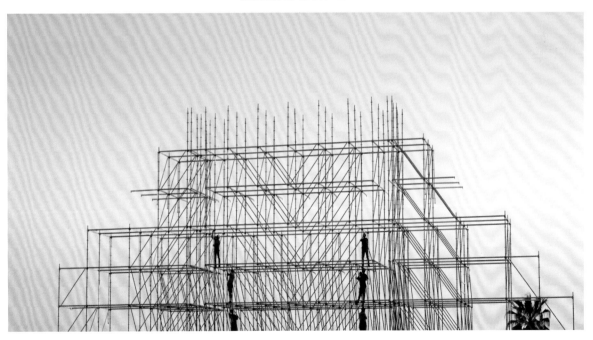

建筑工地的脚手架
Scaffolding on construction sites

钢铁与结构
Steel and Structure

武钢科技图书馆
Science and Technology Library of Wuhan Iron&Steel Co., Ltd.

钢铁小贴士 Tips on Steel

钢是世界上最基础的工程建筑材料。钢应用于我们生活的方方面面：汽车和容器、冰箱和洗衣机、货轮和能源基础设施、医学设备和最新式的卫星。

Steel is the world's most fundamental engineering and construction material. It is used in every aspect of our lives: In cars and cans, refrigerators and washing machines, cargo ships and energy infrastructures, medical equipment and the latest satellites.

钢铁与海洋
Steel and Ocean

钢铁造就了海上巨轮、跨海大桥、海底隧道、防洪工程、人工礁石、海上平台等，钢铁的特性使其成为人类开发和保护海洋的重要材料。
Steel has created marine giant ships, sea-crossing bridges, subsea tunnels, flood control projects, artificial reefs, offshore platforms, and etc. The characteristics of steel make it an important material for human being to develop and protect the ocean.

三亚凤凰游轮码头
Sanya phoenix cruise terminal

钢铁与海洋
Steel and Ocean

人工礁石
Artificial reef

钢渣建造的海底礁石含有高比例的铁和钙等矿物质，可以极大地提高生物量。每吨海底构造物可捕获0.5吨二氧化碳，是名副其实的海底森林。
The submarine reef built by steel slag contains a high proportion of minerals such as iron and calcium, which can greatly increase the biomass. The seabed structure can capture 0.5 tons of carbon dioxide with each ton of artificial reef, which is a veritable seabed forest.

钢铁与海洋
Steel and Ocean

钢制货轮运输了世界上90%的货物,全球有1700万只集装箱大部分由钢铁建造,其优点是结构牢、强度大、水密性好、焊接性高、价格低廉。每天有500至600万只集装箱处在运输途中,每年运输航次数达2亿次。

Steel freighters transport 90% of the world's cargo. Most of the world's 17 million containers are built of steel, which have the advantages of tough structure, high strength, good water tightness, high weldability and low price. About 5 to 6 million containers are in transit every day, with 200 million voyages a year.

大型货轮
Large freighter

钢铁与海洋
Steel and Ocean

巴拿马运河是世界上最具有战略意义的两条人工水道之一。全长81.3千米，水深13~15米不等，河宽150~304米。整个运河的水位高出两大洋26米，设有6座钢制船闸，可以通航76000吨级的轮船。行驶于太平洋和大西洋之间的船只，经过巴拿马运河最长可缩短航程约15000公里。

The Panama Canal is one of the two most strategic artificial waterways in the world. The total length is 81.3 kilometers, the water depth ranges from 13 meters to 15 meters, and the river width ranges from 150 meters to 304 meters. The water level is 26 meters above the ocean and there are 6 steel ship locks. It can sail 76,000-ton ships. Vessels travelling between the Pacific Ocean and the Atlantic Ocean can shorten the journey by approximately 15,000 kilometers through the Panama Canal.

巴拿马运河
Panama canal

钢铁与海洋
Steel and Ocean

世界上许多有海岸线的国家都会建造浮式风力发电设备，这些风力发电所用的材料，80%是钢铁材料。钢铁也是全潮汐开发中使用的核心材料，据估计，潮汐能够满足全球能源需求的20%以上。钢铁为发展可再生能源发挥着至关重要的作用。

Floating wind power plants are built in many countries with coastlines in the world. 80% of the materials used in wind power generation are steel. Steel is also the core material used in the development of all tides power generation. It is estimated that tides can meet more than 20% of global energy demand. Steel plays a vital role in the development of this renewable energy.

海上风电场
Offshore wind farm

钢铁与海洋
Steel and Ocean

钢铁使海上钻井平台坚不可摧,钢铁能够将钻井平台锚定到海床上,为到达困难的储油层提供了重要通道。

Steel makes offshore drilling platforms indestructible, and it can anchor the drilling platform to the seabed, providing an important gateway to the oil reservoirs.

海上钻井平台
Offshore drilling platform

钢铁与海洋
Steel and Ocean

海洋管道是通过密闭的管道在海底连续地输送大量油（气）的管道，是海上油（气）田开发生产系统的主要组成部分，也是目前最快捷、最安全和经济可靠的海上油气运输方式。

The marine pipeline can transport a large amount of oil (gas) on the seabed continuously. It is a major component of the development and production system of offshore oil (gas) fields, and is also the fastest, safest and most economical and reliable method for offshore oil and gas transportation.

海洋管道
Marine pipeline

钢铁与海洋
Steel and Ocean

钢铁小贴士 Tips on Steel

船板 Ship plate

船板是指按船级社建造规范要求生产的用于制造船体结构的热轧钢板材。第一艘焊接钢船大约建造于1940年，此后，轮船几乎全部由焊接钢制造，其价格便宜且重量较轻，逐渐替代了锻铁。今天，大部分巨型轮船都由钢铁建造。钢制轮船运输了世界上90%的货物。

Ship plate refers to the hot rolled steel plate for hull construction, which is produced according to the requirements of Classification Society Construction Code. The first welded steel ship was built around 1940. Since then, the ship has been almost entirely made of welded steel, which is cheaper and lighter, and has gradually replaced wrought iron. Today, most of the giant ships are made of steel. Steel ships transport 90% of the world's cargo.

海洋和谐号
Harmony of the seas

"海洋和谐号"长362米，宽66米，高约64米，仅钢铁部件就超过40万个。游轮有18个甲板，23个游泳池，20个餐厅，载客容量达6000人，重达227000吨，是目前世界上最大的船，比一百年前的"泰坦尼克号"长了近100米。

"Harmony of the seas", 362 meters long, 66 meters wide and 64 meters high, has more than 400,000 steel parts. The cruise ship has 18 decks, 23 swimming pools, 20 restaurants, with carrying capacity of 6,000 people and weighing 227,000 tons. It is the largest ship in the world currently, and nearly 100 meters longer than the Titanic of 100 years ago.

钢铁与海洋
Steel and Ocean

"蛟龙号"载人深潜器是中国首台自主设计、自主集成研制的作业型深海载人潜水器，设计最大下潜深度为7000米，也是目前世界上下潜能力最深的作业型载人潜水器。它的研制成功对我们探索海洋深处提供了设备条件。

The "Jiaolong" is the first self-designed and self-integrated working deep sea manned submersible in China. It has a maximum dive depth of 7,000 meters, making the world's deepest dive ability. Its successful development provides equipment conditions for people to explore the depths of the ocean.

蛟龙号
Jiaolong

钢铁与海洋
Steel and Ocean

泰欧号油轮
TI Europe tanker

　　泰欧号油轮是目前世界上在航的最大油轮，总长380米，船宽68米，载重量为44万吨，满载吃水达24.53米，船速16节。泰欧号已经多次停靠我国油轮码头，为我国输送石油。
　　The TI Europe tanker is currently the largest tanker in the world, with a total length of 380 meters, a ship width of 68 meters, a carrying capacity of 440,000 tons, a full load of 24.53 meters of water, and a speed of 16 knots. It has docked at China's tanker terminal quite a number of times, for transporting oil to China.

钢铁与海洋
Steel and Ocean

港珠澳大桥
Hong Kong-Zhuhai-Macao bridge

港珠澳大桥是连接香港、珠海、澳门的超大型跨海工程，全长55公里，是全世界最长的单体跨海大桥，其中主体工程"海中桥隧"长35.578公里，海底隧道长约6.75公里，沉管隧道的埋深达到46米，是世界最深的。总投资一千亿人民币，用钢量超过42万吨。

The Hong Kong-Zhuhai-Macao Bridge is a super-large cross-sea project connecting Hong Kong, Zhuhai and Macao. It is the longest single sea-crossing bridge in the world with a total length of 55 kilometers. The main project "Haizhong Bridge Tunnel" is 35.578 kilometers long and the submarine tunnel is about 6.75 kilometers long, and the immersed tunnel has a depth of 46 meters, which is the deepest in the world. The total investment is 100 billion yuan, and the amount of steel used exceeds 420,000 tons.

钢铁与海洋
Steel and Ocean

英吉利海峡隧道
The Channel Tunnel

英吉利海峡隧道由两边两条铁路隧道和中间后勤服务隧道共三条长51km的平行隧洞组成，其中海底段的隧洞长度为3×38km，是世界第三长的海底隧道及海底段世界最长的铁路隧道。2013年7月7日，环法冠军克里斯·弗鲁姆从中间的后勤服务通道穿越英吉利海峡，成为了骑自行车通过英吉利海峡隧道的第一人。

The Channel Tunnel consists of three parallel tunnels with 51km lengths, two railway tunnels and an intermediate logistics service tunnel. The length of the tunnel is 3×38km. It is the world's third longest undersea tunnel and the world's longest railway in the seabed. On July 7, 2013, Tour de France champion Chris Frum crossed the Channel Tunnel from the middle of the logistics service channel and became the first person to ride a bicycle through the Channel Tunnel.

钢铁与能源
Steel and Energy

海上炼油厂
Offshore refinery

钢铁对于保障全球的能源供应至关重要。无论是化石燃料、核技术能源，还是可再生能源，钢铁在能源的生产和分配过程中都必不可少。

Steel is crucial to ensuring global energy supply. Whether it is fossil fuel, nuclear technology energy or renewable energy, steel is indispensable in the process of energy production and distribution.

钢铁与能源
Steel and Energy

冰岛的地热发电厂
Iceland's geothermal power plant

地热是冰岛这个寒冷国家的独特能源，现在大约有十分之九的冰岛家庭利用地热能源。
Geothermal is a unique source of energy for Iceland, a cold country, and now about nine out of ten Icelandic households use geothermal energy.

钢铁与能源
Steel and Energy

江厦潮汐发电站
Tidal power station in Jiangxia

江厦潮汐发电站是全球第四大潮汐发电站，这一潮汐发电站能利用潮水涨落双向发电，是世界最先进的潮汐发电工程之一。
Jiangxia's Tidal Power Station is the fourth largest tidal power station in the world. It is one of the most advanced tidal power projects in the world, which can generate electricity by two-way tidal fluctuation.

钢铁与能源
Steel and Energy

钢铁小贴士 Tips on Steel

风电目前占全球可再生能源装机容量的约24%。一台风力涡轮机平均80%由钢组成，这些钢被用于制造塔架、吊舱和转子。一台风力涡轮机平均使用140吨钢。

Wind power currently accounts for approximately 24% of the world's renewable installed capacity. On the average, a wind turbine is comprised of 80% steel, used in the tower, nacelle and rotor. 140 tonnes of steel are required for the average wind turbine.

内蒙古草原上的风电场
Wind power farm on Inner Mongolia grasslands

钢铁是岸上和海上风力涡轮机使用的主要材料。从地基到塔架，从齿轮到外壳，几乎风力涡轮机的每个组件都是由钢铁制成。

Steel is the main material used for wind turbines offshore and onshore. Almost every component of wind turbines, from foundations to towers, gears to shells, are made of steel.

钢铁与能源
Steel and Energy

太阳能
Solar energy

在将太阳能转化成电能或热能上,钢铁起到关键作用。在世界五大太阳能发电厂中,有三个在中国。
Steel plays a key role in converting solar energy into electricity or thermal energy. Three of the world's five largest solar power plants are in China.

钢铁与能源
Steel and Energy

西伯利亚输油管线
Siberian oil pipeline

钢铁小贴士 Tips on Steel
钢管 Steel pipe

钢管按生产方法可分为两大类：无缝钢管和有缝钢管。不仅用于输送流体和粉状固体、交换热能、制造机械零件和容器，它还是一种经济钢材。用钢管制造建筑结构网架、支柱和机械支架，可以减轻重量，节省金属20%~40%，而且可实现工厂化机械化施工。用钢管制造公路桥梁不但可节省钢材、简化施工，而且可大大减少涂保护层的面积，节约投资和维护费用。

Steel pipes can be divided into two categories according to production methods: Seamless steel pipes and seamed steel pipes. It is not only used to transport fluids and powdered solids, exchange heat, make mechanical parts and containers, it is also an economical steel. The use of steel pipes to manufacture building structure grids, pillars and mechanical supports can reduce weight, save 20%~40% of metal, and enable mechanized construction. The use of steel pipes to manufacture highway bridges not only saves steel and simplifies construction, but also greatly reduces the area of the protective coating, and saves investment and maintenance costs.

钢铁与能源
Steel and Energy

输电塔
Transmission line tower

在输电塔和输电电缆中也会使用钢。
Steel is also used in transmission towers and cables.

钢铁与能源
Steel and Energy

新能源汽车充电桩
Charging piles for new energy vehicles

钢铁与智能
Steel and intelligence

酒店服务机器人
Hotel service robot

随着智能时代的来临，钢铁伴随着智能手机、智能汽车、智能家居、智能工厂、智能城市等被赋予了智慧的基因，帮助人类大踏步进入高度智能的世界。

With the advent of the intelligent era, steel is accompanied by smart genes, smart cars, smart homes, smart factories, smart cities and other genes that have been given wisdom, helping humans to step into a highly intelligent world.

钢铁与智能
Steel and Intelligence

远程医疗
Telemedicine

  远程医疗可以使身处偏僻地区和没有良好医疗条件的患者获得良好的诊断和治疗，如农村、山区、野外勘测地、空中、海上等。也可以使全球医学专家同时对在不同空间位置的患者进行会诊，指导手术。
  Telemedicine can provide good diagnosis and treatment for patients in remote areas and without good medical conditions, such as rural areas, mountain areas, wilderness survey sites, air, and sea. It can also enable global medical experts to conduct consultations and guide surgery on patients in different locations.

钢铁与智能
Steel and Intelligence

京东的无人仓库
Jingdong's unmanned warehouse

无人机配送
Drone distribution

京东40000平方米的无人仓库内各种机器人多达上千台，使用了自动立体式存储、3D视觉识别、自动包装、人工智能、物联网等各种前沿技术，兼容并蓄，实现了各种设备、机器、系统之间的高效协同，单日分拣能力达到20万单。京东还在积极布局无人机配送业务。

Jingdong's 40,000 square meters unmanned warehouse has thousands of robots, and uses various cutting-edge technologies such as auto-stereoscopic storage, 3D visual recognition, automatic packaging, artificial intelligence, and internet of things. It achieved efficient synergy between machines and systems, with a single-day sorting capacity of 200,000 units. Jingdong is also actively deploying drone distribution business.

钢铁与智能
Steel and Intelligence

以色列蔬菜大棚
Israeli Greenhouse

滴灌技术　Drip irrigation technology

　　滴灌系统由水源工程、首部枢纽（包括水泵、动力机、过滤器、肥液注入装置、测量控制仪表等）、各级输配水管道和滴头四部分组成，可由电脑任意控制水肥滴灌，使得果蔬味道鲜美。以色列这个缺水国家普及现代化滴灌系统后，水和肥利用率高达90%，农业产出翻了5倍。欧洲市场上的瓜果蔬菜有四成来自以色列，以色列也由此获得"欧洲厨房"的美誉。

　　The drip irrigation system consists of water source engineering, the first hub (including water pump, power machine, filter, fertilizer injection device, measurement and control instrument, etc.), various levels of water distribution pipelines and drip heads. The drip irrigation can be controlled by computer to make fruits and vegetables delicious. After the modern drip irrigation system was popularized in Israel, a water-scarce country, water and fertilizer utilization rates were as high as 90%, and agricultural output was five times higher. 40% of the fruits and vegetables on the European market come from Israel, and Israel has earned the reputation of "European Kitchen".

钢铁与智能
Steel and Intelligence

美国大河钢铁公司
Big River Steel

车间内景
Workshop interior

　　美国大河钢铁得益于先进的炼钢技术及人工智能运营手段，建立智能化、环保型的高品质特种钢制造生态链。已建成的一期实现400人年产量165万吨，预计在二期投产后，可实现600人年产量350万吨钢材的全球最高钢铁生产效率。

Big River Steel has benefited from advanced steelmaking technology and artificial intelligence operation methods to establish an intelligent and environmentally friendly high-quality special steel manufacturing ecological chain. The first phase of the project has achieved an annual output of 1.65 million tons for 400 people. It is expected that after the second phase of production, the world's highest steel production efficiency will be achieved with 350 million tons for 600 people of steel per year.

钢铁与智能
Steel and Intelligence

特斯拉汽车装配车间
Tesla automobile assembly workshop

特斯拉创业团队用IT理念来造汽车,是全球最智能的全自动化汽车生产车间,从原材料加工到成品组装,全由150个机器人完成。

The Tesla startup team uses the IT concept to build cars. It is the world's smartest fully automated car production workshop. From raw material processing to finished product assembly, it all completes by 150 robots.

钢铁小贴士 Tips on Steel

没有钢,就没有电动汽车。硅钢是建造电动汽车发电机和马达的核心材料。
There would be no electric mobility without steel. Electrical steel is an essential material in the construction of generators and motors for electric vehicles.

钢铁与智能
Steel and Intelligence

蔚来无人驾驶
Weilai driverless vehicle

　　蔚来无人驾驶车依托国际领先的交通场景物体识别技术和环境感知技术，实现高精度车辆探测识别、跟踪、距离和速度估计、路面分割、车道线检测，为自动驾驶的智能决策提供依据，目前已得到公路测试牌照，未来将大大降低由于人为操作发生的安全事故。

　　Based on the world's leading traffic scene object recognition technology and environment sensing technology, Weilai driverless vehicle realizes high-precision vehicle detection and recognition, tracking, distance and speed estimation, road segmentation and lane line detection, and provides a basis for intelligent decision-making of automatic driving. At present, road test licenses have been obtained, which will greatly reduce the safety accidents caused by human operation in the future.

钢铁与智能
Steel and Intelligence

青岛无人码头
Unmanned dock in Qingdao

亚洲首个无人码头——青岛码头,有30多辆这样的无人驾驶引导车,载重可达70吨。由于控制精准,停车误差不会超过2厘米,所以在穿梭中不会发生碰撞和摩擦。

Asia's first unmanned terminal, Qingdao Dock, has more than 30 such unmanned guided vehicles with a load capacity of 70 tons. Due to the precise control, the parking error will not exceed 2 cm, so there will be no collision and scrape during the shuttle.

钢铁与智能
Steel and Intelligence

随着人民生活水平的提高，越来越多的家庭购买了汽车，但停车难的问题随之而来。智能立体车库以如何利用有限的土地资源扩大车位量，便捷的存取车为切入点，将传统的地面/地下单层车库改装为多层存车库，采用钢结构，保证安全，充分利用空间，扩展了车位。

With the improvement of people's living standards, more and more families have purchased cars, but parking problems followed. The intelligent stereo garage uses the limited land resources to expand the parking space, and the convenient access to the car is the key point. The traditional ground or underground single-storey garage is converted into a multi-storey garage, which adopts steel structure to ensure safety and make full use of space, and expands parking spaces.

智能立体车库
Intelligent stereo garage

钢铁与智能
Steel and Intelligence

小米平衡车
MIUI balancing vehicle

　　小米生态链是以手机、电视、路由器三大产品线为中心，打造成一个可连接一切终端的大型硬件生态系统。现在已经有包括电饭煲、净水器、平衡车、电动窗帘、扫地机器人等上百个含有钢铁元素的智能硬件，为人们的生活提供安全、舒适、快捷和便利。

　　The MIUI ecological chain is centered on three major product lines of mobile phones, TVs and routers, and it is a large-scale hardware ecosystem that can connect all terminals. Now it has hundreds of intelligent hardwares such as rice cookers, water purifiers, balance cars, electric curtains and sweeping robots to provide people with safe, comfortable, fast and convenient life, all of which are based on steel.

钢铁与循环经济
Steel and Recycling Economy

世界钢铁协会发布的材料中称，钢铁和猫都有九条生命。钢铁不仅可100%循环利用，它的副产品也可以有效利用起来，把经济活动对自然环境的影响降低到尽可能小的程度。

According to the materials published by the World Steel Association, steel and cats both have nine lives. Steel can be recycled 100%, and its by-products can be effectively utilized to reduce the impact of economic activities on the natural environment to a smallest possible extent.

焦油合成塑料
Tar synthetic plastic

钢铁与循环经济
Steel and Recycling Economy

高炉水渣透水砖
Blast furnace slag permeable brick

钢铁与循环经济
Steel and Recycling Economy

人工珊瑚礁
Artificial coral reef

废旧的轮船、车厢等钢铁构件可以下沉海底建造人工礁石，目前全球已建造400余处，成为海底生物的天堂。
The steel scrap from scrapped ships, cars and such kind can be used to build artificial reefs under the sea. At present, more than 400 places have been built around the world, becoming a paradise for underwater life.

钢铁与循环经济
Steel and Recycling Economy

再生钢电池
Recycled steel battery

一种创新的新方法创造了将不锈钢重新用于可持续钾离子电池的可能性。电动汽车数量的增加将需要新的电池生产方法来满足需求。

An innovative approach creates the possibility of reusing stainless steel for sustainable potassium ion batteries. New battery production methods will be required to meet the increase of the number of electric vehicles.

钢铁与循环经济
Steel and Recycling Economy

钢铁小贴士 Tips on Steel

钢是循环经济的基础。通过回收、再利用、再制造和再循环,钢确保资源价值最大化。自1900年以来,已有超过250亿吨废钢得到再循环。这使得铁矿石消耗量减少大约350亿吨,煤炭消耗量减少大约180亿吨。

Steel is fundamental to achieving a recycling economy. It ensures the maximum value of resources through recovery, reuse, remanufacturing and recycling. Over 25 billion tonnes of steel scrap have been recycled to make new steel since 1900. This has reduced iron ore consumption by around 35 billion tonnes, as well as cutting coal consumption by 18 billion tonnes.

废弃汽车再利用
Abandoned car recycling

一辆旧汽车,废钢占80%,有色金属占3%,垃圾占15%,损耗占2%。把废旧汽车零部件和钢铁循环利用起来,能让报废的汽车变得有价值。

In an old car, scrap accounts for 80%, non-ferrous metal accounts for 3%, garbage accounts for 15%, and losses account for 2%. Recycling makes scrapped cars valuable.

钢铁与循环经济
Steel and Recycling Economy

俯视海绵城市
Overlooking sponge city

将自然途径与人工措施相结合，让城市能够像海绵一样下雨时吸水、蓄水、渗水、净水，需要时将蓄存的水释放并加以利用。
Combining natural measures with artificial ones, cities can absorb water, store water, seepage, and clean water like a sponge when it rains, and release and use the stored water when needed.

钢铁与循环经济
Steel and Recycling Economy

废油漆桶
Waste paint bucket

经过油漆桶破碎机的处理进行分离,破碎物在破碎腔内可得到充分而有效的细碎,经过空气回旋分拣系统,把油漆和钢分离开,从而达到废钢再利用。

After being separated by the treatment of the paint bucket crusher, the crushed material can be fully and effectively crushed in the crushing chamber, and the steel can be separated by the air swirl sorting system from paint, achieving scrap steel recycling.

钢铁与循环经济
Steel and Recycling Economy

钢渣肥料的主要有效成分是磷酸四钙$Ca_4(PO_4)_2O$和硅酸钙的固溶体,并含有镁、铁、锰等元素。五氧化二磷含量约12%~18%,是可溶性肥料,可以做酸性土壤的改良剂,以帮助农作物生长。

The main active ingredient in steel slag fertilizer is a solid solution of tetracalcium phosphate $Ca_4(PO_4)_2O$ and calcium silicate, and other elements such as magnesium, iron and manganese. The phosphorus pentoxide content is about 12% to 18%. It is a soluble fertilizer. It can be used as an acid soil improver to help grow crops.

钢渣肥料(钢渣磷肥)
Steel slag phosphatic fertilizer

钢铁与循环经济
Steel and Recycling Economy

沿海钢厂的海水淡化装置，利用海水脱盐生产淡水，可以保障钢厂的用水，还可以提供给沿海居民饮用。
The desalination plant of coastal steel mills uses desalination of seawater to produce fresh water, which can guarantee the water consumption of steel mills and can also be provided to coastal residents for drinking.

海水淡化
Desalination

钢铁与循环经济
Steel and Recycling Economy

废旧轮胎回收
Waste tire recycling

废旧轮胎不再是废弃物,也能成为炼钢厂材料。通过加工处理后生产的炭黑、油品和钢丝等产品具有环保再生、品质高的特点。
Waste tires are no longer waste and can be used as steel mill materials. Products such as carbon black, oil and steel wire produced through processing have the characteristics of environmentally-friendly regeneration and high quality.

钢铁小贴士 Tips on Steel

钢具有无限次的再循环能力,可以反复用于制造汽车、容器和建筑物。零废弃战略、资源的优化使用及钢的超高强度,使钢具备众多可持续发展的优势。
Steel, infinitely recyclable, can be used in cars, cans and buildings over and over again. Zero waste strategies and optimal use of resources, combined with steel's exceptional strength, offer the steel sustainable benefits.

钢铁与生活
Steel and Life

家用电器
Household electric appliances

2017年仅我国家电用钢就超过了两千万吨，为家家户户的生活提高了品质。

In 2017 alone, more than 20 million tons of household appliances steel were used in China, which improved the quality of life for households.

钢铁小贴士 Tips on Steel

家电用钢 Household appliances steel

目前，空调、洗衣机、冰箱、电脑等国内主要家电产品几乎有一半已穿上宝武集团产品的"外衣"。家电板的年供应量目前约达200万吨，包括冷轧产品、热镀（铝）锌产品、电镀锌产品。

At present, almost half of the major domestic appliances such as air conditioners, washing machines, refrigerators, computers and so on have worn the "coat" of Baowu Group products. Nowadays, the annual supply of household appliances boards is about 2 million tons, including cold-rolled products, hot-dip (aluminium) zinc products and electro-galvanized products.

钢铁与生活
Steel and Life

手工艺制作台
Handicraft desk

钢铁与生活
Steel and Life

伦敦大本钟
Big Ben in London

百达翡丽腕表
Patek Philippe watch

无论是重达13.5吨的伦敦大本钟,还是号称"蓝血贵族"的瑞士百达翡丽腕表,准时的背后都有许多个钢铁部件在默默运行。

Whether London's Big Ben, weighing 13.5 tons, or Swiss Patek Philippe's wristwatch, known as the "blue-blooded aristocracy", there are many steel parts running silently behind the timeliness.

钢铁与生活
Steel and Life

烧烤盘
Barbecue tray

不锈钢制作的工具，方便了我们的生活。
Stainless steel tools bring to our living convenience.

钢铁与生活
Steel and Life

钢铁小贴士 Tips on Steel

镀锡板 Tin plate

镀锡板，俗称马口铁，源于澳门（Macao）的译音，当时澳门是马口铁进入中国的主要口岸。目前使用的镀锡板全部为电镀锡（铬）产品，是重要的制罐原材料，用镀锡板制成的罐头盒很容易装运、堆放和储存。其典型用途是用于消费品和工业品的包装。

Tin plate, commonly known as Makou steel, originated from the transliteration of Macau (Macao), which was the main port for tin plate to enter China at that time. The tin plates currently in use are all electroplated tin (chromium) products, which are important canning raw materials. Cans made of tin plate are easy to ship, stack and store. Tin plate is typical used in the packaging of consumer and industrial products.

每年全球要生产2000亿罐食品，钢罐强度高、耐腐蚀，可以更好地保存食品和饮料。

Every year, 200 billion cans of food are produced globally. Steel cans, with high strength and corrosion resistance, can protect food and beverages better.

钢铁与生活
Steel and Life

装修工具
Fitting tools

钢铁与安全
Steel and Safety

钢铁为我们生活的方方面面提供着安全保护。工作中保护脚趾的钢头靴、储存食物的钢制罐头、越来越高的建筑大厦、助我们出行的各种车辆等，钢铁通过不同的途径为我们保障安全。

Steel provides security for all aspects of our lives, such as steel toe boots that protect the toes at work, steel cans that store food, higher and higher buildings, and various vehicles that help us travel.

跳楼机
Drop zone

钢铁与安全
Steel and Safety

亚丁湾护航
Convoy in Gulf of Aden

在亚丁湾附近海域，索马里海盗日益猖獗，严重影响到我国商船及国际组织人道主义物资船舶的安全。根据联合国安理会有关决议，我国海军舰艇编队自2008年到2018年共30批次赴亚丁湾、索马里海域承担护航任务，有效地维护了该地区的海上安全。

In the waters near the Gulf of Aden, Somali pirates have become increasingly rampant, affecting the safety of our merchant ships and international organizations seriously. According to the relevant resolutions of the UN Security Council, Chinese naval fleet formed 30 batches from 2008 to 2018 to go to the Gulf of Aden and the Somali waters to undertake escort missions, effectively maintaining the maritime safety of the region.

钢铁与安全
Steel and Safety

攀岩
Rock climbing

攀岩运动有"岩壁芭蕾""峭壁上的艺术体操"等美称，是极限运动中的一个重要项目。攀岩的时候准备好安全铁锁和绳套尤为重要，这样可以在攀登过程中自我保护。

Rock climbing has the reputation of "rock ballet" and "art gymnastics on the cliff", which is an important project in extreme sports. It is especially important to prepare safety locks and rope sets when climbing, so that you can protect yourself during the climbing process.

钢铁与安全
Steel and Safety

消防栓
Fire hydrant

在城市中各处都备有消防栓、灭火器、喷淋系统和防火门,这些都是钢铁制品,当火灾发生时,它们就会第一时间派上用场,保护我们的安全。

Fire hydrants, fire extinguishers, sprinkler systems and fire doors are available throughout the city. These are steel products that will come in handy to protect our safety when a fire breaks out.

钢铁与安全
Steel and Safety

核电站
Nuclear power plant

安全是核电的生命线，核电站中采用钢铁材料制造部件占整套核电机组部的83%。在这些钢制部件中，制造难度最大的压力容器钢占比最高为14%，反应堆压力容器钢可在高温、高压、物流冲刷和腐蚀的环境下连续工作，且在核电厂整个寿命期都不用更换。

Safety is the lifeline of nuclear power. The use of steel materials in nuclear power plants accounts for 83% of the entire nuclear power unit. Among these steel parts, pressure vessel steel, the most difficult to manufacture, is up to 14%. The reactor pressure vessel steel can work continuously under the conditions of high temperature, high pressure, substances erosion and corrosion, and no replacement during the entire life of the nuclear power plant.

钢铁与安全
Steel and Safety

过山车

Roller coaster

过山车让你刺激的呐喊，别忘了是复杂的钢结构让你娱乐的同时，保护着你的安全。
The roller coaster makes you scream, but don't forget that the complex steel structure keeps you safety while enjoying entertainment.

钢铁与安全
Steel and Safety

城市摄像头
City camera

　　中国已经建成世界上最大的视频监控网——"中国天网",视频镜头超过2000万个,并利用人工智能和大数据进行警务预测,在中国不仅全面普及,而且水平位居世界前列。

　　China has built the world's largest video surveillance network—"China Skynet", with more than 20 million camera lens, and uses artificial intelligence and big data for police action forecasting. It is not only universally popular in China, but also ranks among the best in the world in the network.

钢铁与安全
Steel and Safety

钢铁小贴士 Tips on Steel

压力容器钢 Pressure vessel steel

用于制造石油、化工、石油化工、气体分离和储运气体的压力容器或其他类似设备的钢种,它包括碳钢、碳锰钢、微合金钢、低合金高强度钢以及低温用钢。工作温度一般在-20℃至500℃,个别可达560℃。

Steel, used in pressure vessels or other similar equipment for the manufacture of petroleum, chemical, petrochemical, gas separation and storage of gases, includes carbon steel, carbon manganese steel, microalloyed steel, low alloy high strength steel and low temperature steel. The working temperature generally ranges from -20℃ to 500℃, and the highest can reach to 560℃.

储油罐
Storage tank

我国建成舟山、镇海、大连、兰州、天津及黄岛国家石油储备库共9个国家石油储备基地,储备原油3773万吨。储油罐由压力容器钢制成,工作温度一般在-20~500℃,个别可达560℃,有效保证了石油存储安全。

China has built 9 national oil reserve bases in Zhoushan, Zhenhai, Dalian, Lanzhou, Tianjin and Huangdao, and reserves 37.73 million tons of crude oil. The oil storage tank is made of pressure vessel steel, and the working temperature is generally from -20℃ to 500℃, and the top temperature can reach 560℃, which effectively ensures the safety of petroleum storage.

钢铁与安全
Steel and Safety

钢结构住宅
Steel structure house

　　钢铁是改善住房需求的理想材料。我国是地震高发国家，震区灾后重建的钢结构住宅，不仅起到有效的抗震作用，未来还可以百分之百地回收利用。

　　Steel is the ideal material to improve housing demand. China is a high-risk country with earthquakes. Steel structure houses rebuilt after the earthquake plays an effective role in earthquake resistance, which can be recycled 100% in the future.

钢铁小贴士 Tips on Steel

　　在地震高发地区，钢架结构具有延展性和灵活性的优势，更能抗震。
　　In earthquake prone zones, steel frames have the added advantage of ductility and flexibility to resisit earthquake.

钢铁与安全
Steel and Safety

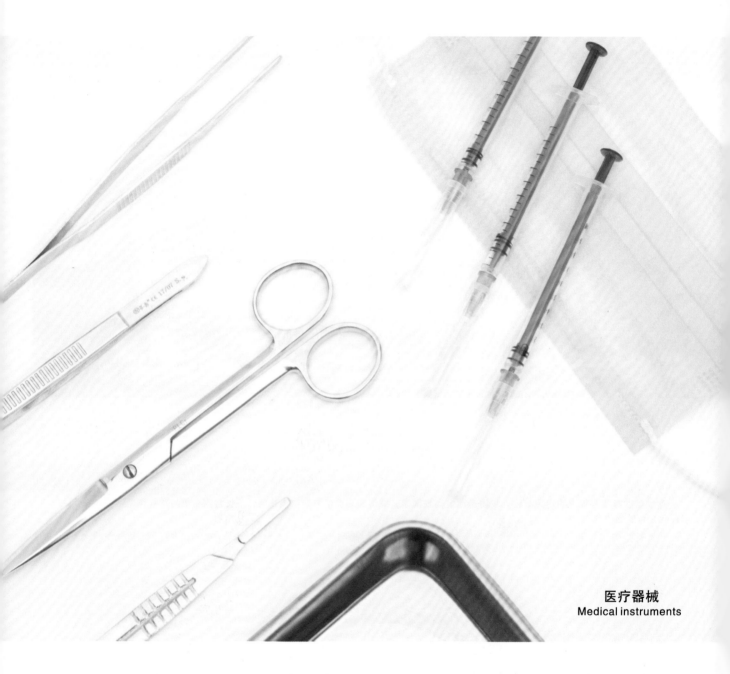

医疗器械
Medical instruments

不锈钢无毒、无化学惰性和无吸收性，可以安全地消毒，而不会腐蚀或降解，这些品质确保它是维持手术室和医院卫生环境安全的重要组成部分。

With non-toxic, chemically inert and non-absorbent properties, stainless steel can be safely sterilized without corrosion or degradation. It is an vital part to maintain the sanitary environment of operating rooms as well as the whole hospitals.

钢铁与安全
Steel and Safety

防浪堤
Breakwater

荷兰北海保护工程
Netherlands North Sea protection works

由于海平面的上升，荷兰部分地区已在海平面以下，由钢铁和混凝土筑起的防浪堤，不仅保护了城市，海鸟也悠闲地在上面休憩。
Due to the rise of sea level, parts of the Netherlands have been below sea level. The breakwaters built of steel and concrete not only protect the city, but seabirds also relax on it.

钢铁与安全
Steel and Safety

灯塔和航标灯
Lighthouse and beacon light

钢制的灯塔和航标灯，在黑夜里发出可视信号，让船员分辨方向，认清航道，保障航行安全。
The steel lighthouse and beacon light emit a visual signal in the dark, allowing the crew to distinguish the direction and recognize the navigation channel to ensure safe navigation.

钢铁与健康
Steel and Health

健身哑铃
Fitness dumbbell

从呱呱落地起，人的健康就离不开钢铁，钢铁制品表面既卫生又易于清洗，这些特性使其成为人们生活中的好朋友。

From the birth, people's health is inseparable from steel. The healthy of steel products is both hygienic and easy to clean, which makes it a good friend in people's lives.

钢铁与健康
Steel and Health

带牙箍的姑娘
Girl with brace

不锈钢的牙箍，帮助我们矫正牙齿，以展示美好的笑容。
Stainless steel dental braces help us correct our teeth to show a beautiful smile.

钢铁与健康
Steel and Health

不锈钢肥皂
Stainless steel soap

不锈钢材料经特殊打磨的磨砂表面与物体摩擦，释放出来的铁离子跟异味分子结合达到祛味的效果。
The stainless steel material is rubbed against the object by a specially polished frosted surface, and the released iron ions combine with the odor molecules to remove peculiar smell.

钢铁与健康
Steel and Health

铁壶
Iron pot

铁壶煮水泡茶，不仅可以软化水质，而且可以释放铁离子，补充人体所需铁质。
Brewing tea with iron pot can not only soften the water, but also release iron ions which are needed by the human body.

钢铁与健康
Steel and Health

公共健身器材
Public fitness equipment

钢制健身器材成为社区的一道风景。耐用、坚固、节能、环保的健身器材帮助人们增强力量，强健体魄。
Steel-made fitness equipments, durable, sturdy, energy-saving, and environmentally-friendly, have become a landscape of the community, helping people strengthen their bodies.

钢铁与健康
Steel and Health

不锈钢餐具
Stainless steel tableware

不锈钢餐具含有大量的Cr（铬）、Ni（镍）合金元素，防止了与食物中的酸碱物质发生反应，也避免了一般铁器生锈后特殊的铁腥味，而且不会摔碎、寿命长。

Stainless steel tableware never breaks up and has a long service life. Containing a large amount of Cr (chromium) and Ni (nickel) alloy elements, it prevents the reaction with acid and alkali substances in food, and also avoids the special iron smell of rusted iron.

钢铁与健康
Steel and Health

校园投币洗衣机
Campus coin-operated washing machine

校园里的公共洗衣机，滚筒及外壳都是钢制。
The drums and the outer casing of the public washing machines on the campus are made of steel.

钢铁与健康
Steel and Health

蔬菜大棚
Greenhouses

用钢结构的骨架制成的蔬菜大棚，形成了一个温室空间，让人们一年四季都能吃上新鲜的蔬菜。
A vegetable greenhouse made of a steel skeleton forms a greenhouse space that allows people to eat fresh vegetables all year round.

钢铁与健康
Steel and Health

咖啡勺
Coffee spoon

钢铁小贴士 Tips on Steel
不锈钢 Stainless steel

　　不锈钢是耐空气、蒸汽、水等弱腐蚀介质和酸、碱、盐等化学侵蚀性介质腐蚀的钢。实际应用中，常将耐弱腐蚀介质腐蚀的钢称为不锈钢，而将耐化学介质腐蚀的钢称为耐酸钢。不锈钢的耐蚀性取决于钢中所含的合金元素。它广泛应用于餐饮厨具、卫生洁具、化工设备、家用电器、食用机械、汽车配件、电子原件、建筑装潢、核电等领域。

　　Stainless steel is a kind of steel which is resistant to weak corrosion medium such as air, steam, water as well as other chemical corrosion medium like acid, alkali and salt. In practical application, steel resistant to weak corrosion medium is often called stainless steel, while steel resistant to chemical corrosion medium is called acid-resistant steel. The corrosion resistance of stainless steel depends on the alloying elements contained in the steel. It is widely used in catering kitchenware, sanitary ware, chemical equipment, household appliances, food machinery, automotive accessories, electronic original parts, building decoration, nuclear power and other fields.

钢铁与健康
Steel and Health

医院的许多医疗设备都是钢制的,医疗级的不锈钢保护着我们的健康。
Many of the medical equipments are made of steel. The medical grade stainless steel protects our health.

核磁共振设备

Nuclear magnetic resonance instrument

钢铁与运动
Steel and Sports

钢铁和运动紧密相连，钢铁不仅支撑了一座座体育场馆，也是许多运动项目的主要材料，如举重、链球、自行车、射击、滑雪等。同时，钢铁也保护着我们运动和欣赏运动时不受伤害。

Steel and sports are closely linked. Steel not only supports stadiums, but also is the main material for many sports, such as weightlifting, hammering, cycling, shooting, skiing, and etc. At the same time, steel also protects us from injuring when sporting and appreciating the sports.

沙滩足球
Beach soccer

钢铁与运动
Steel and Sports

北京国家体育场
Beijing national stadium

　　北京国家体育场（鸟巢），外部钢结构使用了4.2万吨钢材，整个工程总用钢量11万吨，是世界上最大的钢结构建筑。在2008年奥运会开幕式和闭幕式上，给全世界展现了惊艳的美丽。

　　The Beijing National Stadium (Bird's Nest) uses 42,000 tons of steel for the external steel structure and 110,000 tons of steel for the entire project, which is the world's largest steel structure. At the opening and closing ceremonies of the 2008 Olympic Games, it presented amazing beauty to the world.

钢铁小贴士 Tips on Steel

　　2018年，前几大产钢国分别是：中国9.283亿吨，日本1.065亿吨，印度1.043亿吨，美国8670万吨，韩国7250万吨。2018年全球粗钢总量相当于建造43100个北京国家体育馆的用钢量。

　　In 2018, the top steel producing countries were: China 9.283Mt, India 1.065Mt, Japan 1.043 Mt, United States 8670Mt, South Korea 7250Mt. The weight of the crude steel produced in the world in 2018 is equal to the steel required to produce 43100 Beijing National Stadiums.

钢铁与运动
Steel and Sports

跳钢管舞的姑娘们
Dancing girls

钢管舞（Pole dance），是利用钢管为道具的新型舞蹈，受到了不少年轻女孩的青睐。
Pole dance, a new type of dance that uses steel pipes as props, has been favored by many young girls.

钢铁与运动
Steel and Sports

世界一级方程式锦标赛（FIA Formula 1 World Championship），简称F1，与奥运会、世界杯足球赛并称为"世界三大体育盛事"。F1赛车对汽车的速度、操控、安全性都有极高的要求，从这个角度讲，它促进了汽车工业的技术进步。
The FIA Formula 1 World Championship, referred to as F1, is called "the world's three major sports events" together with the Olympic Games and the World Cup. The F1 car has extremely high requirements for the speeding, handling and safety, which promotes technological advancement in the automotive industry.

F1赛车
F1 racing car

钢铁与运动
Steel and Sports

健身房
Gymnasium

随着人们生活水平的提高，健身、运动是现代都市人新的生活方式。"请人吃饭，不如请人出汗"成了现在的流行语。健身房内的大部分器材都是钢制的。

With the improvement of people's living standards, fitness and sports are the new lifestyles of modern urbanites. "It is better to invite people to sweat than to eat" has become the buzzword. Most of the equipment in the gym is made of steel.

钢铁与运动
Steel and Sports

　　冰鞋上的冰刀是由韧性高、延展性好、耐腐蚀性强的不锈钢制成，能够持久保持冰刀的锋利性，为运动员取得好成绩创造条件。雪橇和滑雪板的边缘都用了硬度极高的不锈钢打造。
　　The blades on the skates are made of stainless steel with high toughness, ductility and corrosion resistance. They can maintain the sharpness of the blades for a long time and create conditions for athletes to achieve good results. Skis and snowboards have stainless steel edges with a specific hardness.

滑雪工具
Skiing tool

钢铁与运动
Steel and Sports

这个女孩叫牛钰，10年前的她还是一个年仅11岁的小女孩。然而就在2008年5月12日的那天，坐在教室里和同学一起上课的她，遇上了地震……

10年之后她21岁，踩着钢铁义肢，用一种特殊的方式出现了在了大家的面前——2018汶川马拉松的赛道上。

This girl, named Niuyu, was a little girl of 11 years old 10 years ago. However, on the day of May 12, 2008, she sat in the classroom with her classmates and met an earthquake…

Today, 10 years later, she is 21 years old, stepping on the steel prosthetic, and appearing in front of everyone in a special way – on the track of the 2018 Wenchuan Marathon.

钢铁与运动
Steel and Sports

击剑运动是一项历史悠久的传统体育运动项目，是绅士的运动。1896年在雅典举行的第1届现代奥运会上就设有男子击剑比赛。重剑、花剑、佩剑，都是不锈钢制作的钝剑，比赛和训练时带的面罩也是由金属网制成，用以保护面部不受伤害。

Fencing is a gentleman's sport with a long-established tradition. In 1896, at the 1st Modern Olympic Games in Athens, there was a men's fencing competition. Epee, foil and sabre are blunt swords made of stainless steel. The masks used during competition and training are also made of metal mesh to protect the face from injury.

击剑运动
Fencing

钢铁与艺术
Steel and Art

钢铁良好的适用性，有助于设计师设计出美好的钢铁艺术品。一些钢铁雕塑和钢结构建筑已成为城市的名片。
The good applicability of steel helps designers to design beautiful steel works of art, some of them have become the city cards.

广场上的雕塑
Sculpture on the square

钢铁与艺术
Steel and Art

新加坡滨海花园
Singapore marina garden

滨海花园设计灵感源自新加坡国花"卓锦万黛兰"。花园正中央的几棵巨型擎天大树巍然耸立，分别介于9~16层楼高，是由钢结构打造。夜幕时分，精心设计的灯光秀（Garden Rhapsody）在巨树丛林之间打造出五光十色的奇幻夜空，令人目眩神迷。

The design of the Marina Garden is inspired by the Singapore National Flower "Zhuojin Wandailan". Several giant stalwart trees, between 9 and 16 stories high, standing in the center of the garden, are made of steel. At night, the elaborately designed Light Show (Garden Rhapsody) creates a dazzling night sky between the giant trees.

钢铁与艺术
Steel and Art

鼓浪屿同心锁
Gulangyu concentric lock

铁锁不仅能保护我们的安全,而且也能寓意美好的爱情。
Iron locks can not only protect our safety, but also imply beautiful love.

钢铁与艺术
Steel and Art

商场里的铁艺装饰
Iron art in shopping malls

钢铁与艺术
Steel and Art

阿姆斯特丹·走钢丝的艺术家
Amsterdam · wire walker

武汉欢乐谷
Wuhan happy valley

钢铁与艺术
Steel and Art

石墙上的门扣
Door buckle on stone wall

锈蚀的铁锚
Corroded iron anchor

钢铁与艺术
Steel and Art

《海上生明月》
A bright moon rises at sea

电子探针对耐候钢锈层进行元素分布表征及对钢基中颗粒夹杂进行元素分析时呈现出波浪及圆月状，呈现出一幅"春江潮水连海平，海上明月共潮生"之景象。

The element distribution of weathering steel rust layer was characterized by electron probe, and the element analysis of inclusions in steel matrix showed wave and crescent shape, showing a scene of "Spring river tides at sea level, and bright moon rises with sea tide".

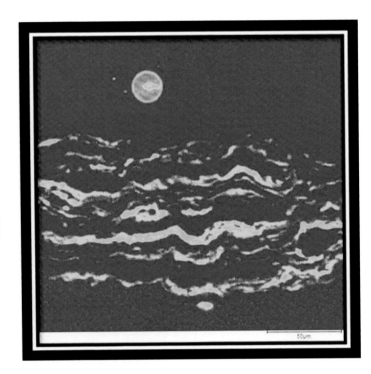

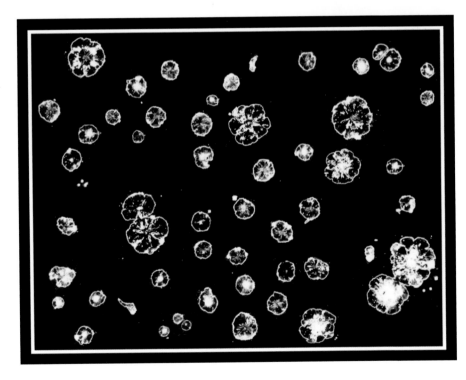

《水晶梅》
Crystal plum

球墨铸铁中的开花状石墨在金相显微镜下的暗场像一朵朵绽放的梅花。

Flowering graphite on dark field in nodular cast iron under metallographic microscope resemble blooming plum blossoms.

钢铁与艺术
Steel and Art

曼德拉雕塑
Mandela sculpture

  位于南非祖鲁-纳塔尔省的纳尔逊·曼德拉纪念碑是为纪念曼德拉被捕50周年于2012年8月4日建成。雕像由高达5~10米的50根金属柱组成，游客只有在35米开外的一点才能辨识出曼德拉的头像。
  The Nelson Mandela Monument in Zulu-Natal, South Africa, was completed on August 4, 2012 to commemorate the 50th anniversary of Mandela's arrest. The statue consists of 50 metal columns up to 5 to 10 meters. Visitors can only recognize Mandela's head at a point 35 meters away.

后 记
POSTSCRIPT

《发现钢铁之美》科普图书，在中国金属学会，湖北省、武汉市科协及中国宝武集团和宝钢股份科协的关怀和指导下，经过编著人员的共同努力，现已编辑完成，这是继《钢铁科普丛书》出版之后武钢科协、湖北省金属学会编著的又一部科普力作，是冶金行业科学普及工作的又一项丰硕成果。

Thanks for the concern and guidance of the Chinese Society for Metals, Hubei Association for Science and Technology, Wuhan Association for Science and Technology, China Baowu Steel Group and Association for Science and Technology of Baoshan Iron&Steel Co., Ltd., "Amazing Steel" has been compiled. This is another popular science book compiled by the Association for Science and Technology of Wuhan Iron&Steel Co., Ltd. after the publication of "the Series of Popular Science of Steel". It is another fruitful achievement in the popularization of science in metallurgical industry.

《发现钢铁之美》编著历时两年，从数千张精美图片中反复甄选素材，采用中英文对照的形式。全书由钢铁与城市、钢铁与工业、钢铁与农业、钢铁与工程、钢铁与艺术等17个系列组成，书中真实展现了钢铁与我们的生活密不可分的联系，以及钢铁为社会进步所做的贡献。

"Amazing Steel" has been edited for two years. Photos in this book have been selected from thousands of pictures carefully, and been compiled in both Chinese and English. There are 17 series, including Steel and Cities, Steel and Industry, Steel and Agriculture, and etc. The book truly shows a tight connection between steel and our life, and tremendous contributions to society of steel.

《发现钢铁之美》全面、客观地反应了钢铁与社会、经济、生活的紧密联系。它的编纂完成得益于相关单位的大力支持；得益于编写组成员认真负责的工作态度。在此，向为本书提供支持帮助的单位和个人表示诚挚的谢意。

"Amazing Steel" reflects the close connection of steel with society, economy and life comprehensively, which has been benefited from the vigorous support of related units, as well as the hard work of the compilation team. Here, we would like to express our sincerely thanks to the units and individuals who provide supports and assistances.

本书收录的摄影作品来源于武钢科协举办的"发现钢铁之美摄影大赛"参赛作品，同时在相关网站获得了部分摄影作品的授权，在此向摄影作者表示感谢！

The photographs included in this book are from the " Photography Contest of Discovering the Beauty of Steel" held by the Association for Science and Technology of Wuhan Iron and Steel Co., Ltd. At the same time, we have obtained authorizations for some other photographs in the photo websites. We would like to express my gratitude to all the photographers.

由于经验不足，水平所限，《发现钢铁之美》还有许多不足之处，诚请读者指正。

It is unavoidable for the deficiencies due to our limited experience and ability. We will be honored to have your valuable suggestions and comments.